INTERNATIONAL P

BANTA
AS
LAYERS

BANTAMS AS LAYERS

J BARNES

Beech Publishing House
Station Yard
Elsted
Midhurst
West Sussex GU29 0JT

First Published 1996

ISBN 1-85736-193-8

Beech Publishing House
Station Yard
Elsted
Midhurst
West Sussex GU29 0JT

CONTENTS

FOREWORD

This is a book which attempts to spread the word that ***smallness can be useful, as well as beautiful;*** bantams, said to originate originally from a region of that name in Java, have been kept for generations, but mainly for showing.

Undoubtedly, many of the *natural* or *ornamental breeds* are primarily exhibition birds, but other, such as the Mediterranean breeds, can be quite good layers. Moreover, they have many advantages over their large sisters – they take up less space, eat much less food, and are able to be kept without difficulty by children, disabled, and elderly people, as well as those who simply wish to provide fresh eggs for the home at low cost, with the minimum of trouble.

This book explains the different types, the breeds to select for laying, their management and many other facets of what can be a fascinating, yet profitable, pastime.

Dedicated to the Memory of

W Powell-Owen

who worked for many years to improve the standards of utility fowl and advocated the utilization of bantams as layers.

Selecting the Layers

On the page opposite the breeds Ancona and Andalusian are likely to be good layers.

The middle row offers Blue-green eggs from the Araucanas, but very few eggs from the Aseel.

The bottom row provides reasonable layers and, with the correct strain of Barnevelders, there should be Dark Brown Eggs.

Ancona
Andalusian
Araucana
(Lavender)
Aseel
(Spangle)
Australorp
Barnevelder
(Double Laced)
Charles Francis

1

BANTAMS
THE BACKGROUND

"Far too many are under the impression that bantams lay very few eggs and small at that, and are therefore not worth the ground space. This is not the case for given a good strain of a "laying" variety, such, for instance, as the Old English Game Black-Reds, eggs will be very plentiful.

If one desires a breed for eggs, he would naturally not take up a variety that was bred solely for its "Fancy" points. The eggs are, of course, smaller than those of (large) hens, but they are just as nutritious and many bantams lay eggs as large as pullet eggs. For invalids bantam eggs are ideal, as they are for children, when a hen's egg would be too large to eat at one meal."

An extract from
Poultry Keeping on Money-Making Lines
W Powell-Owen, FBSA

Silkies, which are good layers, but are frequently broody

THE BACKGROUND

Classification:

Small Poultry:

(a) Natural Bantams

(b) Large Fowl reduced in size by selective breeding.

Brief History

We must accept the fact that the bantam is a miniature of the large fowl or a naturally small fowl. However, some fowl are small anyway so it would be a mistake to assume that smallness always equates to bantams.

Silkies are prime examples, where the **standard size** is 3 to 4 lb (around 1.50 kilo), and yet the Brahma male bantam is around 2 lb. and Orpingtons may be up to 36 oz. But the upper limit of the **large** Brahma is 12 lb. and the Buff Orpington is 10 lb. Originally, in the UK the Silkie was regarded as a light-breed **large** fowl, but in the USA the standard Silkie is a bantam only. In the UK there are now bantam Silkies. A light breed bantam would weigh under 2 kilos and a heavy breed about 2.5 k.

Weight in fowl is important because this determines the food intake. The volume of feathers is also of importance because these have to be maintained and grown after a moult. This is one of the reasons why some fancy fowl are not good layers –

they simply use up too much protein with feather maintenance.

In the past some authorities have referred to bantams as "Lilliputian Breeds" basing the connection on ***Gulliver' Travels*** by Jonathan Swift (1726) where one Lemuel Gulliver landed on an island where the people were six inches tall. Fortunately, the comparison is not valid because on the whole bantams are agreeable creatures and live in harmony even with large fowl, although, cheekily, the bantam cock will be on the lookout for a larger mate should the opportunity present itself. Sometimes, the poultry keeper who is lax in his penning of birds, suddenly finds he has bred some medium size birds when he was expecting large fowl.

The Weights

As noted, the weights of bantams is usually regarded as **25 per cent** (a quarter) of the standard size. This guide is an **approximation** and on analysis it will be found that there are variations in this standard. For example, if Old English Game bantams are taken, with a standard weight of 18 to 26 oz. the large Game is 5 to 6 lb (80 to 96 oz.), it will be seen that the proportion does not work out exactly as indicated. A useful exercise is to take the standard weights for the different breeds and work out the proportion.

Earlier writers looked upon **20** per cent as the standard size compared with the large ***fowl standard***. W F Entwisle writing in 1890 (***Bantams*** published 1892, and still available), stated the rule was one fifth of the standard size. There have been many changes since then and, although the permitted weight has changed upwards, bantams are more natural; they are no longer dwarfs (implying a certain loss of gracefulness), but are diminutive, well proportioned birds, which are beautiful as well as small.

Research has shown that the percentage may vary from around 19 per cent for Cochins and over 28 per cent for Leghorns and Indian Game. There is a dwarfing gene which can produce the bantamized fowl and, when it appears when breeding large fowl, the small bird can be bred back to the parent and the offspring then selected until the bantam finally emerges.

Ability to Lay

The most important consideration is whether the breed and strain selected has the potential to lay. The **breed** is the specific type, possessing characteristics which make it different from other breeds. Thus if we take Leghorns we have a light weight breed which is usually a good layer; but an Indian Game will be quite heavy, slow to develop, and will be an indifferent layer. Within the breed structure there may be different colours and these are known as breed **varieties.**

A further analysis is the way the breed was developed - the gene structure - and whether this includes the potential to lay well. Surprisingly to the amateur, a breed which was quite good at laying may suddenly become poor layers because

Modern Game

Rumpless

Frizzles

Fancy Bantams

Make ideal pets, but not good layers

of a cross introduced to improve show qualities; a cross with an Indian Game to give a broader breast on Old English Game could have disastrous results on the egg production. This final classification is known as the **strain**, and is vital to understand its characteristics when selecting layers.

From this statement it follows that when dealing with show birds, the achievement of some requirements may be contrary to egg production. Such factors are: for instance, a certain type of tail, profuse feathering, an elaborate comb, and keeping to a very small size, all of which are non-utility features. This is not to say that the show requirements are always undesirable, but those which are greatly over-exaggerated often are, and spoil the desired utility features.

The answer is to select those breeds/strains which possess utility features and stick to them. If a brown egg is looked for, then select those bantam breeds which are known to lay brown eggs. There is one breed - the Araucana from Chile which lays green-blue eggs, and this is a novelty in itself.

A Backgarden House.

ADVANTAGES OF BANTAMS

1. Acceptability

In a community are more acceptable than large fowl; cocks' crow is not loud; hens do not make as much noise or get excited.

2. Smaller Space.

Bantams take up less space and eat a smaller amount of food. Many of them lay fewer eggs than large fowl and they are smaller, but the fancier does not seem to mind these facts.

3. Tameness and Ease of Handling.

They become very tame and can be kept as pets, suitable for children.

4. Easier for Show Preparation.

For preparing for shows and transporting there, they are much more manageable than large fowl, which are four times the size.

5. More Variety.

A greater variety and many more breeds can be kept in a relatively small space.

6. Very Attractive.

Bantams are small, attractive and often quite beautiful.

7. Suitable for Study.

For schools and similar establishments wishing to study biology, zoology, incubation, and genetics they are easy to keep.

8. Easy to Rear.

Generally, after the first few weeks (when extra care is needed with the chicks hatched) they are quite strong, and generally trouble free.

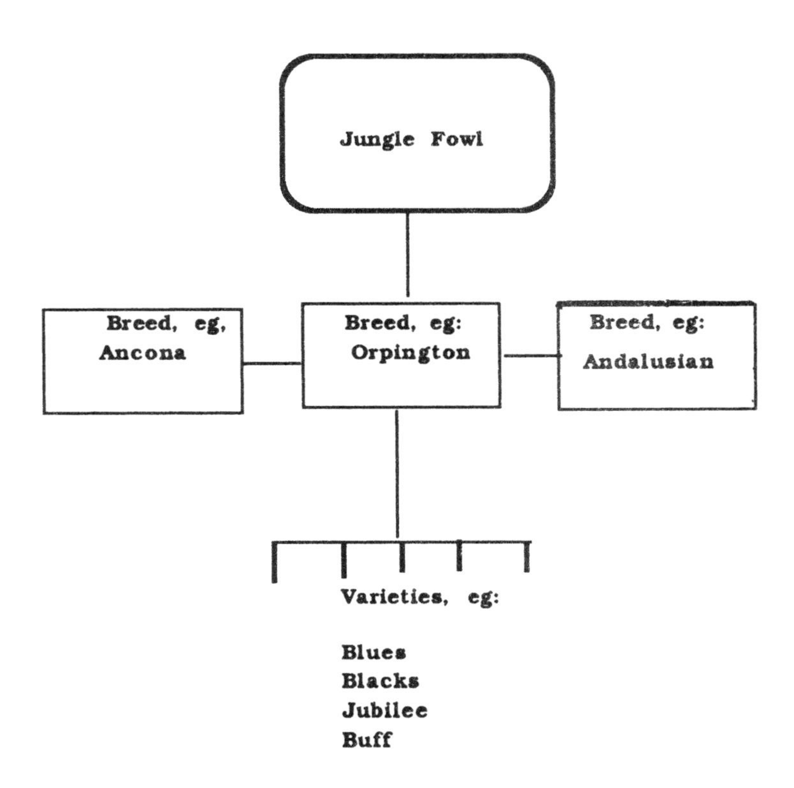

Thus it will be appreciated that the breed is divided into different varieties based primarily on colour.

Development of Bantams

Came from Red Jungle Fowl and then developed in different areas or was created from different crosses.

9. Less Likely to Damage Garden.
Bantams kept at the bottom of a garden - even running on a lawn - do not do any significant damage, whereas large fowl, must have much more space and their scratching can be harmful.

10. Suitable for Showing & Profitable.
For the fancier who wins top prizes the stock fetches high prices and if he or she wishes there is plenty of activity at poultry shows and in judging.

DISADVANTAGES OF BANTAMS

1. Smaller Eggs.
Although true, it takes less food to produce them and those with a small appetite do not mind, eg; children.

2. Lay Fewer Eggs than Large
This is true, especially in winter months, but with careful selection and correct conditions, such as heat and light, paricularly the latter, a laying strain can be developed.

3. Ornamental or Fancy Points are Against Egg Production.
This is true so if a laying strain is to be developed select layers and/or utility type birds.

A Movable House.

BANTAMS ARE IDEAL AS A HOBBY

Keeping bantams is very much the realm of the hobbyist, although there are a few bantam-poultry farmers who manage to make a profit from the activity. The aim usually is to enjoy keeping and breeding birds which are useful as well as attractive. For most people who keep bantams, the ultimate aim is to win prizes at shows and thereby improve the bantams kept. Each year, by improving the stock by careful and skilful breeding, the **fancier** derives great satisfaction from his hobby.

Those seeking eggs must concentrate on the utility aspect and if showing is to be undertaken then *suitable* breeds must be kept.

EGGS

The eggs produced by bantams may be distinguished in many ways:

1. **Colour,** eg; Dark Brown, Brown, Light Brown, Cream, Speckled, and White; the Araucana breed lays greeny-blue eggs. The colour makes no difference to the food content of the egg, but in the UK there is now a marked preference for brown eggs, yet the best layers tend to produce white or cream eggs.
2. **Size.** Bantams lay eggs in the region of 1.50oz. Although some breeds lay larger eggs, bearing in mind the size of the hen, this should not be encouraged.
3. **Quality of Shell;** this should be smooth with a strong shell (no obvious pores) and should exhibit a good bloom.
4. **Egg Content; *yolk*** should be a deep yellow and bright; ***white*** (albumen) of egg as white a possible with no blood spots; the shape should be ovoid and quite regular.
5. **Freshness;** this is indicated by a small 'air-space' at broad end.

Since the purpose or ***objective*** is to produce eggs every effort must be made to give the hens the conditions to achieve maximum production of top quality eggs.

Note: The 1.50 oz (42.50g.) is given as a guide only and as a maximum for bantam eggs. However, many bantam pullets will not lay such large eggs. Nearer one ounce may be more likely at first and some may never get to the maximum size.

2

CLASSIFICATION OF BREEDS

"Bantams, we suppose, always have had, and always will have their admirers, and, we think, in increasing numbers as the years roll on; for, regarded from any standpoint one may choose, they have much to be said in their favour."

Bantams
William Flamank Entwisle
The "Father" of the Bantam Movement

BRAHMA BANTAMS

DARK COCK. DARK HEN. PAIR OF LIGHT BRAHMAS

Some Early Bantams produced by W F Entwisle in the 1880s

CLASSIFICATION

Basic Requirements

There are many many of classifying bantams, for example, by the following:

1. **Type,**
2. **Size,**
3. **Whether natural bantams,**
4. **Country of origin,**

and so on.

For the purpose of this book the approach is influenced by the ability to lay eggs. There are more than 60 breeds of poultry and many of these are in bantam form or have been at some time.

If bred so they are recognized as a distinct breed they are listed in an official description known as a ***Standard***. If two breeds are bred together they become cross breeds and cannot be shown, except as utility breeds, where crosses are allowed.

The usual categories are as follows:

1. **True Bantams**

2. **Ornamental Bantams**

3. **Laying Breeds**

4. **Utility or Heavy Breeds**

This classification serves as a guide on which breeds to select for egg laying purposes. The laying breeds of **large fowl** may lay up to 300 eggs in a season, but these kind of results were obtained after many years of domestication and selection. Bantams do not come into this category, but with selection will improve until quite profitable.

True Bantams

These are naturally small and continue to remain this size. Thus the following qualify:

1. **Belgians**
2. **Booted**
3. **Dutch**
4. **Japanese**
5. **Nankins**
6. **Pekins (Cochins)**
7. **Rosecombs**
8. **Sebrights**
9. **Tuzos**
10. **Watermael**

The Frizzle is sometimes put into this category, but it does have a large equivalent. The sprightly Sebright bantam was developed by Sir John Sebright around 200 years ago, but there seems little doubt that it is a ***true*** bantam.

In addition, there are some breeds that are now true bantams, although not classified as such. If we take Old English Game bantams, now far removed from the large fowl, these are now true bantams. Modern Game come into a similar category and breed as bantams without any proper link back to large Modern Game which are very rare anyway. Similar arguments could also be found for other breeds which were not produced from the large fowl as such, but usually the term is used for those which are natural bantams from the start.

The top layers are not to be found in this group, although OEG bantams are quite good layers and so are some of the others in the Spring and Summer.

Japanese

Very old breed with large head, short legs and high tail - far removed from the Jungle Fowl

Further True Bantam

These are not suitable for developing a Laying Strain

Booted **Pekins**

Sebright

Ornamental & True Bantams

Kept by enthusiastic fanciers for showing

Ornamental Bantams

Ornamental bantams are Fancy fowl, possessing one or more characteristics which make them unusual, or exceptionally beautiful. Thus we have:

1. **Belgians**, which are quite ornate with rare colours, feathered legs and beards.
2. **Japanese,** with very short legs, a large head and tall tail.
3. **Pekins** which have short, feathered legs, and are ball-like in shape.
4. **Polands,** which have very large crests.
5. **Rosecombs,** with elaborate comb and long tail.
6. **Sebrights,** which are beautifully laced and have hen feathering so that cock and hen are similar.

Others may appear in this category, and some could certainly be looked upon as more ornamental than normal. The Silkie might well be looked upon as an ornamental, being very unusual with fur-like covering instead of feathers. Since the Pekin is included, why not the Brahma, which is very similar and at one point was looked upon as ornamental, although many breeders disagreed with the description.

Some of the others in the True Bantam Group could also be looked upon as ornamental, such as Nankins and Dutch. The inclusion of Frizzles seems a natural choice. Rumpless bantams are unusual and therefore could qualify.

The Poultry Club decide the classification so much depends on the opinion at the time. Breeds which could well be considered are Hamburghs, Houdans, Old English Pheasant Fowl, Redcap, Spanish, Sultans, Sumatra Game, and Yokohamas. Since a category may change, readers are advised to seek the correct, up to date classification at a particular time.

The fact of the matter is that, although these are very attractive as show birds, they are not likely to be in the top layer category. The fancy points are the guide for breeding and these are often in opposition to the ***utility aspects.***

Silver Campines

An exotic looking bird which may be a reasonable layer.

The Laying Breeds

The laying breeds are primarily the light breeds which originated in Italy and Spain and are known as the Mediterranean Breeds.

As large fowl, their development has been over hundreds of years. The bantams are of more recent origin and should be miniatures of the large fowl, even to the detail of being **non-broody.**

The large fowl lay over 200 eggs and many lay 300 or more. Moreover, because of their body size, they are relatively small eaters.

The breeds which come into this category are:

1. **Ancona**
2. **Andalusian**
3. **Leghorn**
4. **Minorca**

In shape they all follow a similar pattern; the males have an upright tail and the females tend to have a slim tail; the body is "egg shaped", and the legs medium to long. The feathers are close fitting, almost 'hard feathered' a description accorded to Game breeds. Because they are non-sitters no eggs are lost in a broody period.

Bantams are not such good layers, but many strains provide excellent results for bantams. Even for some of the rarer colours, such as Pile Leghorns, claims have been made for very acceptable levels of production. At present this stands at around 160 to 200 eggs per annum ***for laying strains,*** but with selection and the correct conditions, could be improved upon.

Breeds outside the Mediterranean Group which can do well are Rhode Island Reds, Sussex, Plymouth Rocks and Wyandottes, but the correct ***laying strain*** must be the aim not the show strains. These are utility types, which are heavier in build than the Mediterranean breeds.

Ancona Leghorn

Top Layers

Utility and Heavy Breeds

The breeds mentioned in the preceding paragraph come into this category. They are heavier and generally come broody, which can be a disadvantage. The Sussex are the most popular of these all-round types.

When we consider the real heavy weights, such as the hard feathered breeds of Malays and Aseel, and their 'offspring' the Indian Game, these must be discounted because they lay very few eggs. For all practical purposes, therefore, such breeds are best left out of the analysis of what might be suitable. They are 'non-runners' right from the start.

Interesting Breed

These are poor layers and therefore are kept as show birds

Minorca

Orpington Sussex

A Mixture of Layers

The soft feathered types such as Brahmas, Langshans, Dorkings and Faverolles are also unsuitable as layers, although some strains may be reasonable.

Classification of the Main Breeds

The final analysis of the likely layers are listed below. Remember, though, that some breeds have many varieties; Leghorns, Wyandottes, and Sussex are examples of where there is a multitude of colours to suit most tastes. On the other hand, Anconas and Rhode Island Reds have just one variety each.

A **recap** of the possible breeds is:

Laying Light Breeds (laying white eggs)

Ancona
Andalusian
Leghorn
Minorca
Spanish (used to be good layers, but now doubtful - excessive ear lobes)

One of the factors to consider with the Mediterranean breeds is temperament. Some strains, not kept in close captivity and tame, may be 'flighty' and may not be suitable for a small garden. However, once established in domesticity, with a little common sense, are quite suitable, because with patience they become tame.

Utility including Heavy Breeds
(lay tinted/light brown eggs)

Croad Langshan *
Hamburgh
Old English Game*
Orpington*
Plymouth Rock*
Redcap
Rhode Island Red*
Scots Grey
Sussex*
Wyandotte*

Dark Brown Eggers

Barnevelders* (not all strains lay dark brown eggs)
Marans*
Welsummers

The heavy breeds are generally taken as the "sitters", but this classification is rather wide because some breeds which appear as if they may be sitters are not, but appear as light breeds. Old English Game are sitters and yet are a light breed.

Bantams follow the same classification as large fowl, but do not always behave in exactly the same way because of the different crosses used to bantamize a breed.

***Classified as *sitters* and therefore may be counted on for hatching and rearing chicks, although there would be a loss in egg production.**

3

SELECTION OF THE LAYERS

"Laying like other qualities is hereditary, but good layers do not always produce layers as good as themselves; and for a long time this was a great puzzle to breeders, until it was discovered that a certain quality for heavy laying acts exactly as though it were sex linked, which at once threw a new light on the whole subject."

Now what do we mean by a moderately good layer?

"If a bird lays no eggs in her first winter, and does not start her life's work until March, she is definitely a bad layer. If, on the other hand, she lays over 30 eggs by the end of February, she is a good layer. If she lays under 30 eggs during this period, she is a moderate layer."

Both extracts from:
Poultry Keeping Made Easier
N W Joergens (now out of print)

Hamburgh

Scots Grey

Wyandotte

Some Further Likely Layers

SELECTION

Breed is First Guide

The **breed** is the first guide to whether a bird is likely to be a moderate or good layer. Once a laying breed has been selected then try to establish whether the particular **strain** is likely to perform at an acceptable level.

Because of unknown factors a strain may not be a good layer, even though from a laying breed. This means that the laying strains must be free from the 'bad' genes of non-laying strains, which were introduced to improve show points. Thus breeds like Orpingtons, bred originally for utility features, have been spoilt by creating a very 'feathery' breed.; similarly, whilst Plymouth Rocks are noted layers in large, the bantam variety is said to have evolved from crosses with Scots Greys, possibly a strain that was a poor layer.

Another factor which has worked against the development of laying strains of bantams has been the adherence to the small size required for some breeds of show bantam. Hatching late in the year, restricting food, breeding from very small parents and other factors have all contributed to the idea that bantams are not layers. Some experts have stated that larger sizes would allow a fuller egg potential to be realized.

If a strain of laying bantams is to be developed then an understanding of the criteria to be used for selection should be understood.

Factors to Consider

The main criteria to be used as a basis for selection may be viewed under the following headings:

1. **Physical Characteristics**
2. **Number of Eggs Laid**
3. **Times When Eggs Laid**
4. **Progeny Testing, also known as the Nucleus System.**

Physical Characteristics

The outward physical characteristics are the 'sign-posts' of health and ability to lay. Features to look for are as follows:

1. Comb and Face

The comb and face should be bright red, although not very 'beefy'; eyes should be prominent and alert; face should be smooth.

2. Activity

A laying hen is a busy hen. On the other hand, a hen which sits around and mopes is suspect. If this continues, unless moulting, most likely she should be culled.

3. Plumage

A healthy bird will have a good covering of feathers which are not broken or dirty. In the case of the latter, where feathers are not shiny and healthy, this is a sure sign that something is amiss.

Feathers missing across the back may mean the cock is running with too few hens, so he is

Barred Plymouth Rock

Silver Wyandotte **Houdan**

Breeds Which May also be Good Layers

treading them too frequently. This probably means that he needs more hens. If there is no intention of breeding, there will be no need to keep a male bird.

The problem may be caused by mite eating the feathers or by feather pecking. Both are curable; by dusting with anti-mite powder or spray, and providing a dustbath. The feather picking or pecking is usually a sign of boredom or a lack of protein in the diet; get plenty of scratching material, fresh at least once per week, grass clippings, leaves, and diggings from clearing up the garden, and the hens will be too busy to resort to vices. Always ensure a fresh supply of water for a shortage can cause many problems.

Number of Eggs Laid

A simple record of the number of eggs laid will be essential. However, good management should spot those which are 'workers', and this fact is backed by the written record on a **hen** or a **pen** basis; rings on each bird help the process, although not vital when only a few birds are kept.

Methods of noting the layers are by handling and spotting the signs:

1. Weight & Condition

A healthy bird is firm of body, with glossy feathers, and a healthy look. If a bird feels very heavy, with obvious fat, it is not likely to be a layer.

19--	LAYING RECORD			WEEKLY
Date	**Hen or Pen**	**No. Laid**	**Size**	**Notes**
21/8	AB	6	A	WATCH SHELLS

Total by week, month and quarter. Note any hens not laying and for how long.

Record of Eggs Produced

2. Examination of Vent

This should be full and moist; healthy with no faeces on the short feathers below. Look out for mite eggs in this area and remove them, adding a smear of sulphur ointment to prevent recurrence. Width of Pelvic Bone Space indicates **Capacity** (see drawing p. 37).

3. Signs of Sexual Desire

If a hen is picked up she may crouch as if expecting to be mated with the cock. This is a sign of a healthy hen. The worker is likely to be aggressive with other hens.

4. Trap Nesting

Special nest boxes may be used which keep the hen on the nest until she is released, and the egg laid is then recorded. There should be adequate nest boxes and regular release and inspection.

A target should be set and any hens not achieving this level should not be bred from. Around 100 eggs may be an acceptable minimum target.

Times When Eggs Laid

As noted at the beginning of this chapter, a good layer will produce eggs in the Winter months, the number determining whether she is a **moderate** or **good** layer. If eggs are produced ***only*** from March to October, she is a **poor layer.**

The number of times a hen becomes broody is also important; the time she takes to get through the moult, and any 'breaks' she has during the year which reduce the total eggs laid. A healthy hen gets over the difficult periods like the

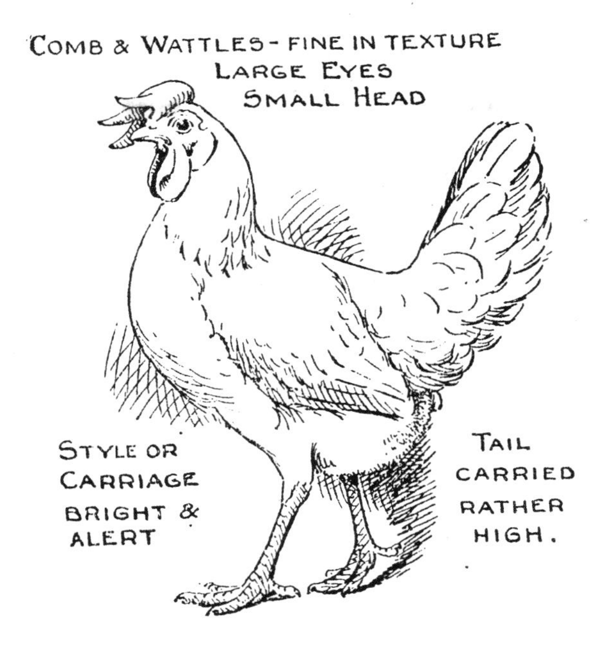

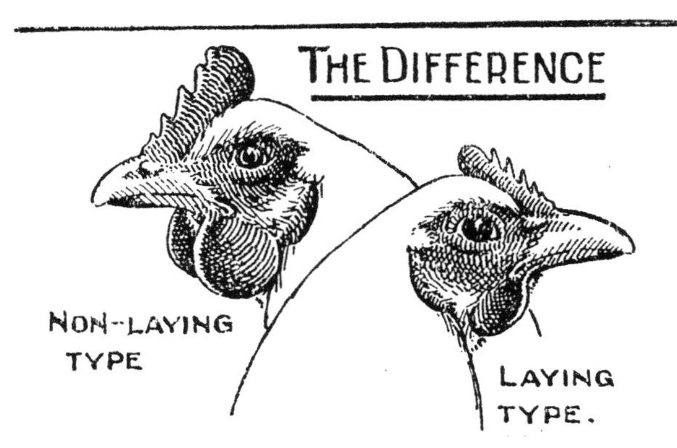

The Laying type Hen

See also page 72.

moult quite quickly, but the unproductive hen waits for a long time before becoming active again

Progeny Testing

This is a method used by geneticists when building up a strain of layers for commercial purposes. It involves the development of **super layers** on **both sides** of the family; ie, even the males should be from parents which produce super layers.

A 'Family' is taken as a basis for the tests and an average of the eggs laid is recorded over a period. Only the families which excel in laying are bred from, using a male which has been bred from a male and hen with the correct background. As noted, some form of inbreeding and line breeding (breeding within the family) will be essential for this ***fixes*** the desired characteristics.

This approach, although producing very good results, ***does take a long time***. Some projects take upwards of 20 years. It also involves rejecting any families which fail to achieve the target set, even though some of the hens may be excellent layers. ***A small breeder will probably stick to the testing of individuals and rely on records kept.***

The measure used for these commercial tests is to take into account all the likely factors and then try to maximize egg production by varying the factors, using different birds. Important factors are: Age when laying started and weight of egg; rate of production; death or illness rate, and such aspects as feather density, size of body, and so on. The measure employed is called the **Hen-housed egg production** and obtains an average of eggs laid from dividing **total laid** by number of hens in the test **at the start**, thus calculating the -esult of non-producers such as those which died.

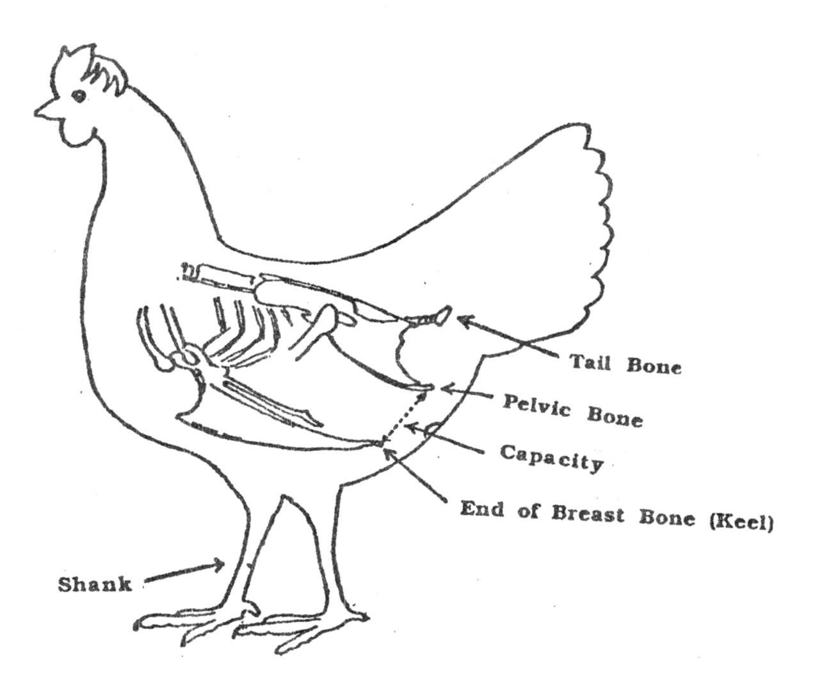

Outline of Good Layer

Pelvic bone should be wide for laying and she should have no physical deformities.

CONCLUSION

Which ever method is used it will be appreciated that the main objective is to select, keep and breed from layers. A standard must be fixed which is just below the best hen kept. This may be, say, 120 eggs per annum, which will be made up from two elements:

1. October to March, say, 30 eggs

2. April to September, say, 90 eggs.

These figures are given to illustrate the main principles; a good strain may lay around 200 eggs.

Any hens which lay below a minimum of, say, 60 eggs should be culled.

At the beginning the target may be set lower than an acceptable level. From this stage it will be essential to ***breed from known layers from a cockerel which is related***. Usually stock would be kept for two seasons and then replaced, although some strains may be worth while for a third year.

The light breeds will not generally come broody so if breeding is to be attempted a breed such as Silkies or Sussex must also be kept; or, alternatively, a small incubator may be used.

4

HOUSING

"Practically any design of house and run will do to accommodate the miniature"

W Powell-Owen

A WELL DESIGNED HOUSE FOR BANTAMS

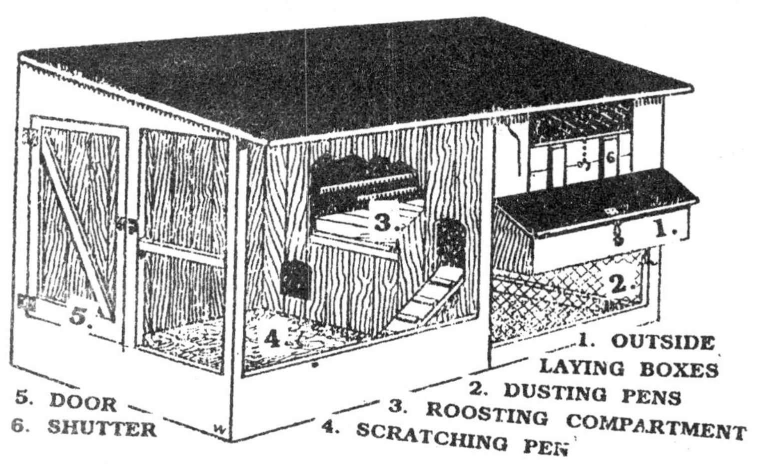

Possible Houses for Bantams

REQUIREMENTS

Utility bantams are quite hardy and will thrive under any normal conditions, whether in an enclosed run or living free range in an orchard or field. The main danger comes from predators such as the fox and therefore it is important to keep a watch for any dangers; above all see they are locked indoors each night.

The various matters requiring attention under ***management*** are as follows:

1. **Accommodation of various types**
2. **Feeding**
3. **Breeding**

These are covered in that order, in the next three chapters.

ACCOMMODATION

With bantams it is necessary to think in terms of winter and summer housing. In summer they can have an outside run or have access to the paddock or yard, preferably where they can scratch for food and eat green food such as grass and chickweed. No special protection is required, but a roosting shed which is dry and free from damp is essential.

When winter comes the bantams may still be let out in the elements, but they are better in a covered run with leaves, peat moss, or wood shavings in which they can scratch. Moreover, although not the best of layers in winter, in the right surroundings they will still lay and lead a normal active life. This is much better than having to cope with rain, sleet and snow which they do not really like.

Essentials in the house or run are the following:

1. Perches

Perches which are not too smooth so the birds can grip easily and fixed above the ground so they can fly up. Do not have them too thick or their toes will be spoilt by trying to grip. If possible, have a stepping effect by having a series of perches. Generally they will keep flying up until on the highest perch.

2. Hoppers/Fountains

The birds must have a supply of fresh water and once each week give the container a thorough clean. The food hopper can contain pellets for ad lib feeding.

3. Nest Boxes

Nest boxes should be provided so that the hens lay inside, lined with hay, straw or shavings. They must be kept clean and the material changed regularly. If an egg gets broken, remove the lining and wash out the nestbox with hot water which includes washing up liquid and disinfectant.

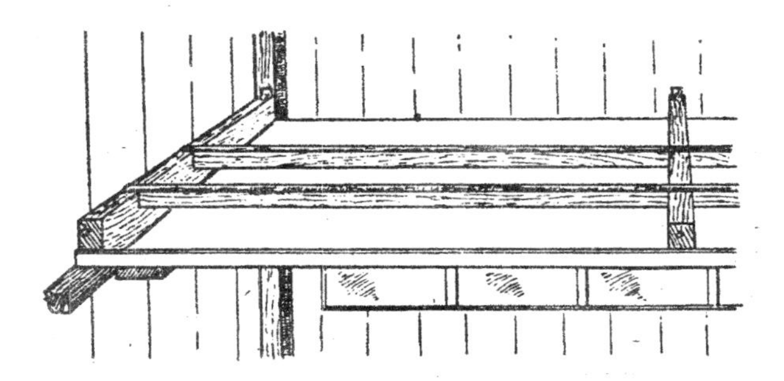

Droppings Board

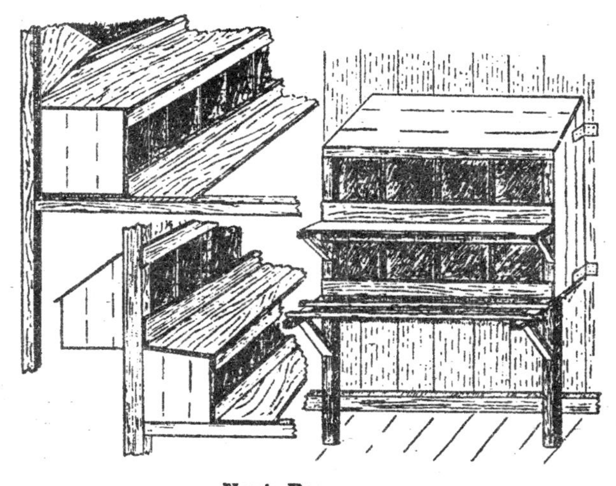

Nest Boxes

Vital Accessories in A Shed

Try to plan sheds so they can accommodate all necessary equipment and can be managed easily.

Small Sheds and Runs

A wide variety of sheds and runs are available or can be made by a poultry keeper with a reasonable flair for carpentry. Many designs are possible to fit the space available.*

Possiblilities are:

1. **House and Run Combined.**
2. **House and Separate Run - this should be covered over or closed off in winter.**
3. **Sussex Ark or similar movable 'house'.**

There has been recognition for many years that for winter eggs light must be provided and the shed, whilst adequately ventilated, should also be well insulated so that the birds feel warm enough to lay. The lighting can be arranged on a time switch or simply by turning on the light each evening at dusk and then turning it off when a **total** of 12 hours light has been given, ie; make up the difference with artificial light.

In designing a hen house it is essential to watch out on economies of management. Keep a **dustbin** or similar container with lid inside each shed to store layers' pellets so feeding is easy. Make sure there is provision for greens by providing some form of **green's rack.**

*** See *Poultry Houses & Appliances* - A DIY Guide available from the publisher.**

A **dust bath** should also be available, filled with sand, cinders and fine earth, which the hen scatters on her to kill mite.

Slatted Floor which allows droppings to fall into a compartment where the manure dries when it can be removed for use in the garden.

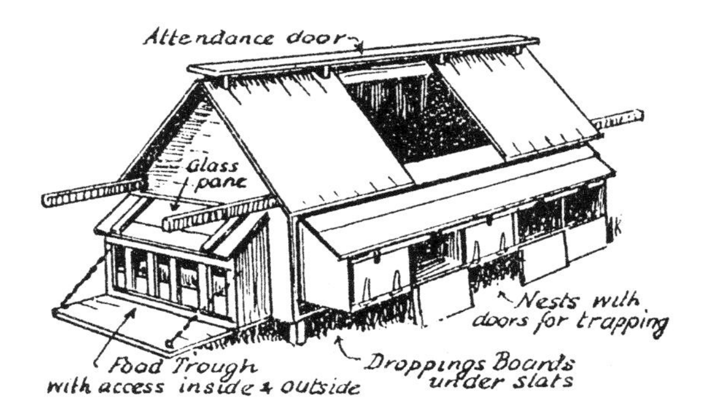

Laying Ark with many features as indicated. Note poles at each end for ease of moving to fresh ground.

Typical Small Sheds

Using A Large Shed

For those with limited space or for winter quarters a garden shed is ideal. A shed 8 ft. X 10 ft. can be divided into two compartments with a door between and two trios can be kept, one in each section. However, be sure to have hardboard along the bottom of the partition because bantam cocks like 'to have a go' and therefore they are better separated.

Also include a platform shelf about 2 ft. wide across the back and about 5 ft from the ground. This acts as a **droppings board** and a nest box can also be included. The height gives the birds exercise as they fly up to reach the food and perches.

The **floor** should ideally be concrete or very strong wood (tarred) to keep out rats and mice. This should then be covered with garden soil and leaves and other materials, added to form deep litter for scratching.

The fancier who desires to keep many birds will require a large shed with a corridor up the middle and compartments on each side with separate doors. Trap doors will then lead out into separate runs so there is ample space for breeding pens. In addition, if showing, or when spare cockerels are kept, or there is a health problem, a Penning Room or shed will be essential for training and/or getting birds into condition.

5

FEEDING

For Laying and especially in Winter

"It was simply the result of providing the layers with comfortable quarters, where they were well protected from rain and winds, and plenty of dry litter in which to scratch, and were given generous allowances of warm food and food which produced warmth. Such, one might say, is what pullets require today – comfortable quarters and sound food for steady production in winter."

From ***Poultry Breeding and Management***

William W. Broomhead
An expert on poultry in the early days of commercial poultry farming.

Water Fountain

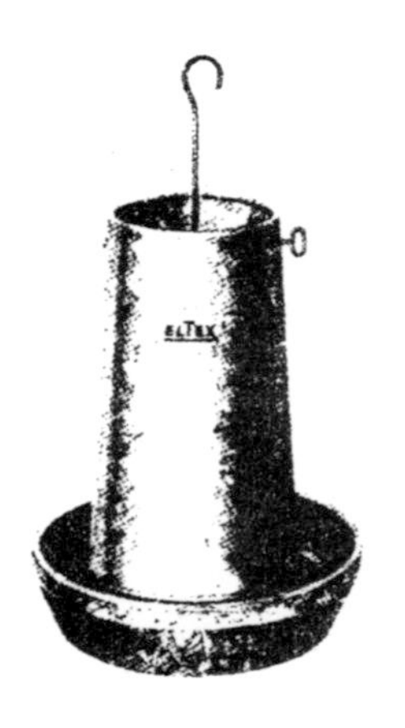

Food Hopper

Food & Water Utensils

FEEDING

Keep them hungry, but feed them well' should be the principle for bantams. This means they are given adequate food according to the season and whether the hens are laying or not. Overfeeding is just as dangerous as inadequate feeding so care should be taken not to give too much food to hens which are not 'working' – being active and laying.

The easy way is to have a hopper full of poultry pellets; this is known as ***ad lib feeding*** and birds can thrive quite well on it. Mash (dry powder) can also be used, but birds find difficulty in swallowing the powder and it gets in the water troughs. In winter it can be used to make a crumbly mash, including a limited amount of household scraps, and the hens will gobble this up. What is needed is variety and presented in such a way that the birds are eager to eat and they are prepared to scratch and exercise not simply sit on a perch all day and eat to capacity morning and evening.

A menu can take many forms, but it must be practical. Pellets of the **small type** for layers can be given in the morning by scattering in the litter. For those who are away early a hopper can be used for birds to help themselves, but the ration for the day should be watched. Some feed mixed corn in the evening, but not all agree for layers because it reduces the protein intake, so essential for laying.

If corn is fed and if the birds eat a good handful each you will know they are not eating too many pellets. If they are sluggish and leave the corn, then only a limited quantity of pellets should be left in the hopper.

Some fanciers advocate a handful of corn or two in the morning and moistened mash in the evening, but this is a matter of preference on the part of the person involved.

Layers' pellets are a **balanced food** and therefore must not be 'diluted' too much or some valuable part will be lost. ***There is an optimum level of protein, amino acids, grit, minerals and so on.*** If household scraps are fed, and these can act as an appetiser last thing in the evening, it must be appreciated, that if too many are given, the birds may not lay as well. One failing is that the eggs may have weak shells because there is no grit in the scraps. A proper level of grit is essential to avoid thin shells which are very undesirable.

Suitable Household Scraps

Edible household scraps to a limit which does not dilute the food value below the requirements of a laying bird can help to economize. Such things as boiled potatoes (not raw), meat and fish (small pieces), cooked vegetables, beans, bacon scraps and other items may be fed. Bread, especially the wholemeal type, soaked in old milk or water, and then squeezed out is very welcome because it stimulates the appetite.

Greens Rack which avoids greens getting soiled is hooked on to wall where birds can reach it.

Feeding Greenstuff & Root Crops to Bantams

Water, Grit & Other Essentials

Water is vital because without it, even for a day the birds will suffer. If for any reason birds are deprived of water - usually due to a water fountain leaking - the combs of the birds will turn a smutty colour and the birds will not lay. In fact, if left short of water, hens may resort to eating their own eggs, a habit which is very hard to break once established. Remember that an egg is about 80 per cent water so without a constant supply of fresh water laying will be restricted.

Grit, both soluble (limestone and oyster shell) and insoluble (small flints) should be available. The former is for calcium and the latter is for grinding up the food in the gizzard. Both aspects are essential.

Greenstuff such as lawn clippings, leaves, chickweed, dandelions, and other wild, edible weeds should be given on a regular basis. If the birds are running outside they will pick their own.

Moulting Time

At moulting time give the birds extra protein to help with feathering, Cod liver oil may be added to the food and extra protein in the form of fish meal, if available, may be a supplementary feed. It is essential to get birds through the moult quickly and get them back into lay. The good layer will soon grow a new coat of feathers, whereas the "wasters" do not recover until well into Spring.

6

BREEDING FOR SUCCESSION

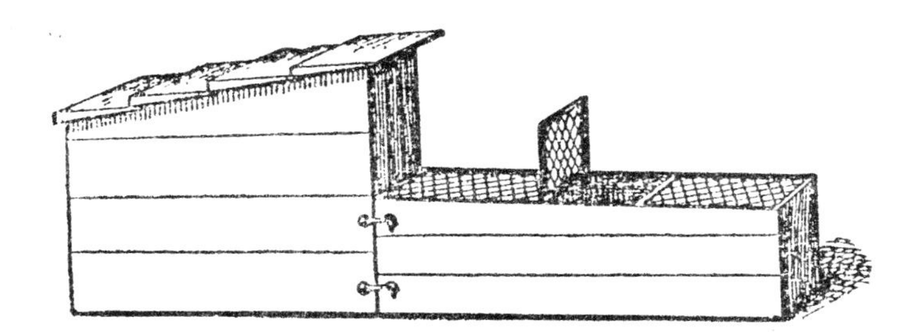

Outdoor Coop for Broody & Chicks

Note the run is 'tied-in' to the Coop, otherwise a fox or dog may push it over. The floor shold have a wire floor to keep out predators such as rats which may burrow.

Select A Reliable Broody

Many heavy breeds are suitable, but can be clumsy on the small bantam eggs. A Silkie or Silkie cross is excellent provided featheres are not tangled.

BREEDING

Breeding is the most enjoyable part of bantam keeping. It presents a challenge and there is always the problem arising which makes the whole process so interesting.

The reader is advised to understand the **aim** or **objective** and in this case it is to maximize laying. For ensuring there is succession to replace old birds, this must be done in a way which leads to improvement of results by the application of scientific method in the breeding pen.

Start the plannning process in the Autumn and put the pair or trio of birds together about November when they have moulted and are getting into condition. Select them carefully along the lines suggested earlier. In summary form this means breeding from birds ***which have an excellent record for laying and a male from a related family which also excel in laying.*** Discard any poor performers right at the start.

Once in their separate pens make them as comfortable as possible and introduce more layer's pellets and a variety of titbits; a little bread soaked in milk seems to encourage the appetite. If early eggs are desired, then have a light on in the evenings which will stimulate the laying and give more time for activity and feeding. Make sure all the birds are in good condition and for early fertility use a young cock. The older bird may not be fertile until later in the season.

Yearling hens (no older early in the season) should be used and use the older, tried hens in another pen later on. However, do not introduce new hens into an established breeding pen be- cause fighting will occur between hens. If any changes are to be made then select an entirely new house and run and start off from scratch with the new breeding trio.

During the breeding season keep the birds active and feed layer's pellets of the small size; let them on to a grass run if possible because this really gets them into top condition. However, if the weather is frosty, or wet and very inclement the birds are better kept inside where they will thrive quite well on deep litter. There should be good light from a window and adequate fresh air. Removable shutters are a sound idea for these allow the worst conditions to be kept out and can be removed when the weather gets better.

Using A Broody Hen

The bantams can be allowed to sit, but a better idea is to use a special broody which is small, yet will cover a dozen eggs. A Silkie or Silkie cross is ideal; a pure Silkie is excellent provided her soft feathers at the front are re- moved because chicks get trapped in them.

Do not use a very heavy breed like an Indian Game, Nor an excitable breed such as large Old English Game because these are too rough on the small bantam chicks early on.

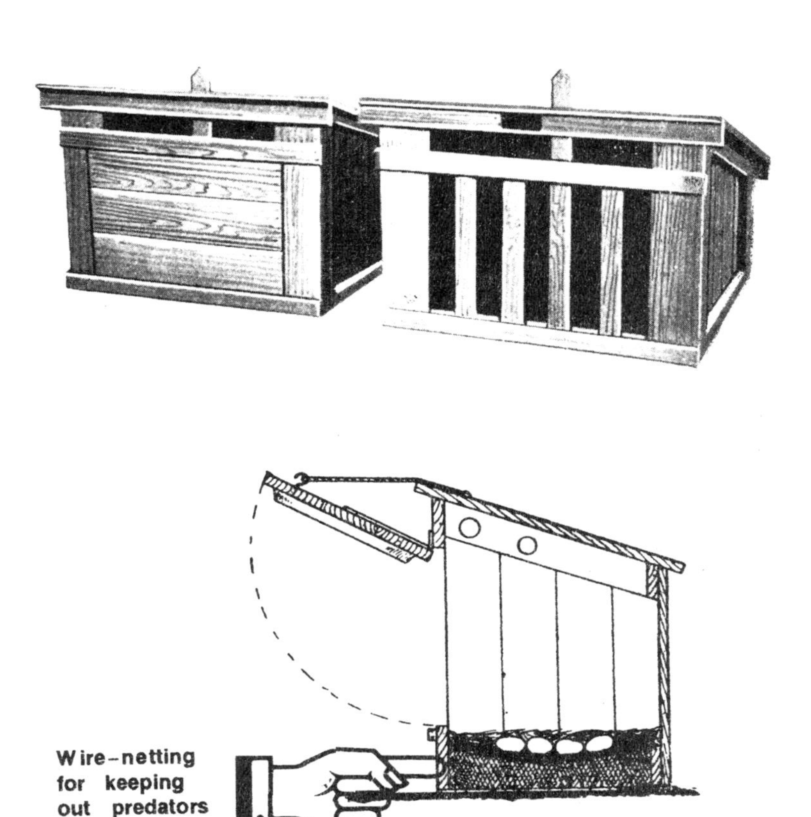

Turf underneath
the Hay or Straw

SITTING NEST WITH FRONT SHUTTER RAISED.

Broody Coop for a Sitting Hen

She should have privacy and let off the nest once per day and fed with hard corn. The base of the nest should be basin-shaped, made of a turf, and lined with straw.

An Incubator

A small incubator is an alternative, ***but is not really worthwhile unless at least 30 eggs are to be set at regular intervals.*** Remember the chicks have to be reared so hatching them alone is only part of the problem. Early in the season an incubator may be the only method available, but remember it only becomes a viable proposition if the procedures are properly organized.*

The essentials for incubation and rearing are as follows:

1. Daily Egg Collection

Collect eggs daily and store them in a cool place in trays or boxes containing shavings. If dirty when collected wash very carefully in water to which has been added a mild disinfectant to sterilize the shell.

2. Identify The Eggs

Mark the eggs with a code for the pen/birds and the date. If to be incubated put an **X** and **Y** on opposite sides of the egg to aid with the turning process; it shows which eggs have been turned, which should be done twice daily to prevent the yolk sticking to the membrane. Damaged, mis-shaped or very dirty eggs should not be incubated, and all eggs set should be fresh.

A Newly Hatched Chick
Should be quite dry before being moved

A Typical Small Incubator
There are many types; avoid flimsy, plastic incubators that will not stand up to cleaning.

3. Prepare the Incubator

After cleaning and sterilizing, pre-heat the incubator and make sure it is warmed up before loading with eggs. These should be no more than 7 days old. The temperature will be in the region of 102 deg. F with the thermometer just above an egg.

4. Candling

The eggs should be tested (candled) at 7 and 14 days. Any that are clear or addled should be discarded.

5. Hatching Time

Expect the eggs to hatch at 19 to 20 days and from the 17th day do not turn or interfere in any way. The moisture is usually increased just before this time.

6. Removal of Chicks

Once hatched, move the chicks to a brooder or broody hen which has been sitting for 10 days or more. Place the chicks under her and gradually move the eggs under her into the incubator. Watch carefully to make sure she takes to them (move quickly if she pecks them), but do nothing more for 2 days except to feed the broody. If a brooder is used (possibly infra-red lamp) pre-heat to 95 deg. F. and then drop by 5 degrees each week by raising the lamp.

7. First Feeding

Feed chick crumbs initially and at 14 days introduce broken corn and chick-weed. After about 6 weeks introduce grower's pellets and continue with the corn. Once fully feathered the chicks can be put in an outside run and they will grow very quickly. Make sure the ground is fresh to avoid diseases. For water use a chick-size water fount and change the water every day.

Infra-Red Lamp & Chicks

Special Points To Watch

When breeding bantams there are one or two matters to watch:

1. Selection of Broody

If a broody hen is to be used try to avoid hens which are large and clumsy or those with feathered legs. On the latter a cross Silkie is better than a pure bred because of the fluffy feathers and the feathered legs which may cause problems such as scaly legs (caused by a tiny mite) which can be passed to the chicks. If it does occur, treatment with sulphur ointment is advisable or some paint on paraffin (make sure this does not touch the skin because it burns), but this can be a dangerous remedy.

2. Watch for Coccidiosis

Chicks that are mopy and appear to be having problems with growing wing feathers (they show very unevenly) are probably suffering from coccidiosis which is a disease of the intestines. Fortunately, the standard chick crumbs contain a ***coccidiostat*** which will prevent too many losses. If there are a number of cases a drug should be introduced into the water (obtainable from a farm shop or by prescription from your vet.)

3. Isolation of Sick Birds

If one of the birds in the breeding pen becomes ill then isolate and try to determine the problem. Feeding on soft food may be the answer, but do not attempt to breed from the breeding pen until the matter is cleared up.

4. Eggs not Hatching

If eggs are clear (ie, not fertile) it may be the cock which is too old or too young. Early in the season a cock may be out of condition, but will improve later. Sometimes old hens are too fat or bully a cockerel so the eggs fail to be fertile.

5. Breed At the Correct Time

Do not be in a mad rush to breed early in the season. Often early chicks catch the autumn shows, but those hatched a few weeks later may progress just as fast; better weather, especially sunshine, can make a tremendous difference to fertility and growth of chicks.

6. Check Regularly for Mite

Watch stock birds to ensure there is no mite on them. Examine around the vent and remove any dirty feathers and rub with sulphur ointment. Provide a dust bath so birds can dust themselves in dry soil or soil and fine ash. A box under cover, filled with the soil is all that is needed. If feathers become affected with feather mite (feathers eaten away) try bathing the bird in a solution of warm water and fairly strong disinfectant, but make sure this does not get into the eyes. In addition, spray with an insecticide, but, again, watch the bird's eyes and also ensure they do not breathe in the fine spray.

7. Separate Males & Females

At about 3 months of age it will be better for males and females to be separated because they thrive better that way and it avoids the fights and squabbles that may take place. The heavier breed cockerels may be worth fattening, but if not dispose of them because the eat a lot of food.

8. Intensive Rearing

Bantams can be reared intensively, but there should be direct light, other than through glass, which implies mild weather conditions. Cod Liver Oil added to the food (a few drops per day and vitamins - Abidec) can help the process and avoids rickets which may appear if the chicks do not get direct sunshine.

9. Discouraging the Broody Hen

If a hen becomes broody and is not required for sitting she can be discouraged by being removed from the nest box where she continues to sit and

The Broody Breaking Coop

This discourages broody hens and avoids loss of production

clucks when removed. Removal to a new shed will probably be sufficient and she will be back in lay in about 10 days, but if left weeks of production may be lost. Some breeders use a Broody Breaking Coop, which is an open-fronted cage with a wire or slatted floor. Continue to feed and water her whilst she continues to be broody.

HATCHING CHART.	
Sitters.	**Non-Sitters.**
The breeds named below are "sitters," and lay tinted or dark brown eggs.	The breeds named below are non-sitters, and lay white shelled eggs.
Hatch out in March or April.	**Hatch out in April or May.**
Wyandottes.	Anconas.
Faverolles.	Andalusians.
Indian Game.	Campines.
Croad Langshans.	Hamburghs.
Rhode Island Reds.	La Bresse.
Sussex.	Leghorns.
Orpingtons.	Minorcas.
Plymouth Rocks.	Buttercups.

Further Reading

Books are available on theses subject areas:

Natural Incubation & Rearing

Artificial Incubation & Rearing

Both by Dr J Batty

7

MAKING A START

FEATURES.	GOOD LAYER.	POOR LAYER.
Head . . .	Well balanced, flat on top and broad.	Long and narrow, or rising sharply in front and falling behind.
Eyes . . .	Full, bright, prominent, set well up in face.	Small, sunken, dull, set low and far back in skull.
Face . . .	Lean, good colour, fine texture.	Fat, coarse-skinned, or pale and thin.
Comb and Wattles	Fully grown, red and warm, fine silky texture.	Shrunken, dry and dull.
Body . . .	Broad back, deep breast, straight, medium length breast bone.	Narrow weedy type, or coarse type with round thick bones.
Shanks . .	Thin and flat with smooth scales.	Round and fat with coarse scales.
Feathers . .	Tight and well worn, late and quick moulting.	New clean feathers, loosely carried, early slow moulting.
Skin . . .	Soft, silky and pliable.	Coarse, dry and thick with hard layer of fat beneath.
Pelvic bones .	Prominent and pliable.	Thick, stiff or covered with fat.
Distance between pelvic bones and breast bone .	If laying, 4–5 fingers breadth between pelvic and breast bones, 2–4 fingers between 2 pelvic bones themselves.	1–2 fingers breadth between pelvic and breast bones, pelvic bones close together.
Vent . . .	If laying, large, moist and pale.	Small, contracted, round.
Yellow colour (in yellow - fleshed breeds) .	Entirely absent in vent, eyelid and beak and shanks	Present in vent, eye, beak and shanks.

Features indicating Good & Poor Layers

Note: Pelvic measurements given relate to large fowl so bantams would be about 35% of those given, possibly less. The important principle is that the space is big enough for ease of laying the 1.50oz eggs.

PURCHASING STOCK

If starting in poultry keeping, or going over to layers, it will be necessary to look around for laying stock which have a proven record of laying. As noted, show birds may not be good layers, simply because they have been bred for "Show Points". Obviously, a laying breed must be sought, but of the utility type.

When purchasing make enquiries from the breeder and ask for information on laying abilities. This may not be readily available because some breeders simply collect the eggs and keep no records; if this is the case, the total number laid will have to be approximated from what information is available, but be prepared to improve on the laying potential by selective breeding. If necessary, purchase hatching eggs, but this is not a wise move unless experienced.

Under no circumstances should "wasters" be accepted in the belief that they will improve with good feeding. These are birds which eat plenty do not produce, simply because the potential to lay is not present. The characteristics which should be expected in the laying type of hen have been enumerated as follows:*

1. Fine comb, long snaky head, and narrow skull.

2. A very red and prominent eye; the more the eye stands out from the head the better.

---

*'*Selecting Poultry for Laying: The Inheritance of Fecundity in Fowls*, **Oscar Smart.**

3. A long body sloping very gradually towards the tail; the tail itself being carried almost but not quite erect.

4. The breast-bone very short, and the abdomen (which should be covered with the shortest possible down) well developed.

5. Legs rather above than below the average in length (but not too long); very fine and extremely pale in colour. The legs should be set wide apart.

6. Toe nails extremely short.

7. Pelvis bones wide apart, not less than 2.50 inches.*

8. Cartilage soft.

It is the combination of these eight characteristics which comprise the laying type - not any few of them by themselves.

The "fancy breeds" such as Brahmas, Cochins, Malays, Aseel, and the short backed, heavily feathered Orpingtons fail on a number of counts and therefore are poor layers.

There are some birds which appear to qualify, but do not lay as well as indicated by the condition. This does not mean that the ***general theory*** is incorrect, but simply that there are some other, possibly unknown factors, from an alien cross which are restricting the laying capabilities. Size is an important factor, because usually the large birds do not lay well because they

*** Bantams would not expect to be quite as wide because the egg is smaller. However, the pelvic bones should be quite open, indicating a laying bird (Compare with a bird not laying when the bones will be tightly closed)**

tend to be lethargic, although this is not the full story.*

The safest approach is to start off with the known laying strains and improve upon them.

Leghorns

Faverolles

See Oscar Smart, (ibid)

HEALTHY BIRDS ESSENTIAL*

If birds are to lay they must be healthy which is usually the norm for those which are kept in dry, well ventilated sheds, preferably with free range in the Spring and Summer months and sheltered accommodation, with lighting in the Winter.

There are also bad habits which must be guarded against if birds are to be laying at maximum. These are as follows:

1. **Feather Pecking.**
2. **Egg Eating.**
3. **Soft Shells.**

Diseases

The diseases encountered in a small flock, on free range or semi-free range, are very small. Those that do occur are relatively minor and with sound management can be spotted at an early age.

1. Fleas & Red Mite

Changing litter regularly or using a deep litter system will keep fleas at bay. Painting the nest boxes and ends of perches with paraffin or spraying with an insecticide will control the parasites.

2. Lice & Feather Mite

These appear on the birds quite regularly if not controlled. Dusting with poultry insect powder (or spraying) will control; dust baths and the placing of sulphur

*** For those wishing to study an advanced book on poultry diseases see *Poultry Diseases Under Modern Mananagement*, G S Coutts, Elsted.**

ointment around the vent will help to keep the mite at bay. If there are dried faeces on the feathers below the vent, or mite eggs have been laid, remove the feathers and rub in a small amount of sulphur ointment. The proprietory brands of insect killer such as Vapona can help to keep the sheds free from flies and mite.

Scaly Leg

This unsightly condition, appearing as raised and roughened scales on the legs, is caused by tiny parasites which live under the skin. The problem is caused by dirty, neglected runs, and is a sign of bad management. Sulphur ointment mixed with around 10 per cent medicinal tar (obtainable from a chemists) will quickly remove the irritation. Significantly birds which are washed regularly for shows do not appear to be affected, which is an indication of the preventative measure required.

Bumble Foot

One of the methods of keeping birds under a modern system is the ***Aviary Method***, where birds have a series is floors or "shelves" within a buildind on to which they can fly. This is an excellent system, but birds flying up high may land badly or land on a stone and an injury occurs. This turns into an abcess and the foot becomes very swollen. The foot should be examined and any glass, thorn or other foreign body sticking in the foot should be removed and the pus allowed to drain; the wound is then sterilized with Dettol or a similar medication.

The wound should be cleaned, washed and the treatment repeated until better. If necessary the bird should be placed in a separate run until the trouble has cleared.

Paralysis

Generally there is no cure for paralysis, but there is a vaccine available against Marek's disease and if a breed is subject to this form of paralysis the Herpes vaccine should be used at a day old. Some strains of Silkies are prone to the problem. This should not be confused with leg weakness which may occur when chicks are reared on wooden floors and/or are fed on food which is low in protein.

Colds & Roup

The birds may get bad colds which lead to blocked sinuses and eyes sometimes close from the swelling. Lack of ventilation and damp conditions can cause the infection; muddy runs and other adverse conditions also contribute to the problem. Antibiotics are the answer and conditions must be improved.

Other Problems

Other problems may occur, but not usually, provided the stock is well managed, and any ailments that appear are deal with promptly. Remember any sick birds will not usually lay so do not delay, and in severe cases, kill the bird affected and burn or bury the carcase.

Medical Chest

If a number of birds are to be kept there should be the means of separating any sick birds – usually in what is known as a 'Penning Room'. This would contain separate cages or compartments, with wire fronts, so any bird to be isolated and treated can be dealt with.

A number of basic ointments would usually be kept in a suitable box. Typical contents would be as follows:

Sulphur Ointment

As noted, this is an excellent remedy for dealing with scaly leg, mite, small scratches or sores or a variety of other problems.

Disinfectant

This may be in the form of powder and in liquid form. It should be used for disinfecting the house, but also for cleaning wounds or similar purposes.

Vaseline

Suitable for rubbing on comb and legs as a protection.

Washing up Liquid

When the birds are washed, which is usual if showing, but is also a good remedy for keeping birds free from mite and other problems. A mild disinfestant is put in the warm water.

Bandages & Plasters

These are for dealing with wounds.

Dubbing Scissors

Essential for Game fanciers, but also if a cock gets frost bite on his comb.

These and possibly others would be essential. Drugs may also be kept for worming and for treating colds, but care must be taken to ensure that they are not out of date and they are not misused, because, as stated, healthy birds do not usually require much attention.

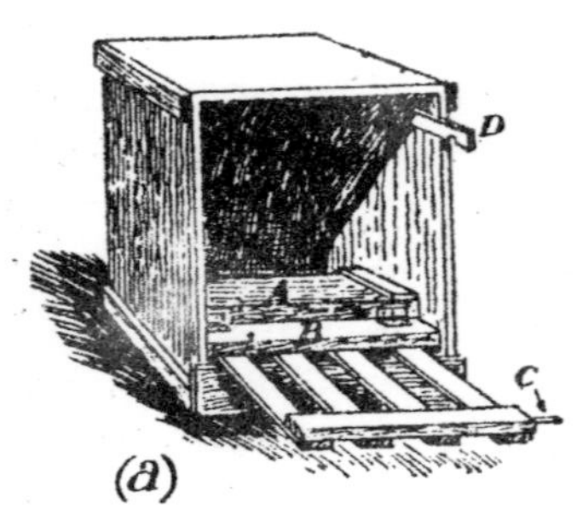

A trap-nest. (*a*) Bar A is pivoted at the back and hinged to B, which is pivoted near the front. To reach the nest, the hen has (*b*) to step on to A, so causing the grid to fly up and spike C to engage with catch D.

USEFUL TOOLS

One of the great advantages of poultry-keeping is that the chores involved in it can be performed at any time; morning, evening, week-ends or other convenient period will suffice. Many business people keep poultry, vicars have been stalwart figures in the Poultry Fancy and others, from all walks of life, manage to enjoy this delightful pastime. Establishing some sort of routine, even though flexible, is essential. Layers' pellets and a gravity fed hopper is a great boon and so is the drinking fountain, which allow food and drink to be available at all times.

The chores of cleaning out, regular collecting of eggs, checking that all is well are still essential and having appropriate tools and equipment hanging in the shed or Penning Room can help to allow the small jobs to be done at convenient times. A list of such tools is now given:

1. **Scraper** for removing manure from droppings board.
2. **Hand brush** for dusting ledges and nest boxes.
3. **Small Shovel** for removal of faeces.
4. **Bucket** for manure.
5. **Spade and Fork** for digging and turning over runs.
6. **Wheelbarrow** for transporting weeds, soil, etc. into runs.

COMMON PROBLEMS

Problems which may arise are

1. **Feather Pecking.**
2. **Egg Eating.**
3. **Soft Shells.**

Feather Pecking.

This vicious trait can occur by accident such as when a quill feather accidently falls and other birds grab it and this develops a taste for blood. Once it starts, especially if the birds are in close confinement, it progresses into a very bad habit.

Giving more space and increasing animal protein may effect a cure. Spraying with water in which a disinfectant has been added may discourage the vice.

Commercial poultry farmers im the past used to 'debeak' the birds, chopping off the end of the beak with a special cutter. This is a nasty practice and is not to be encouraged. Another solution is the use of specially made 'specs', but, again, is more for the commercial people. It should not be necessary for the small scale poultry keeper.

Egg Eating

A soft shelled egg may start the practice. Accordingly, the answer is to ensure the eggs have strong shells with a good 'bloom'. In severe cases an egg may be filled with mustard, or a special 'roll-away' nest box may be built so the eggs roll underneath the false floor. Always collect eggs at regular intervals, thus taking precautions against a likely offender.

Soft Shells

Lack of grit, or a hen which is too fat, or one which cannot for some reason produce the shell, are all possibilities. Make sure grit is always available; sprinkle grit powder into mixed mash, or even feed by hand so that the hen is known to have been given grit; usually the problem can be eliminated, especially if the birds are allowed plenty of exercise which reduces the weight and enables them to pick up grit from the earth.

Over Feeding

A major problem with hens is that they tend to *accumulate* fat internally (due to over eating and lack of exercise) and this can cause problems. Control the food fed so that this problem does not arise.

Poultry are very wasteful with food and therefore strict control is essential; leaving food about also encourages rats. Years ago I used to visit one of the last of the professional poultry showmen, who prepared birds for shows, bred top class specimens for sale at high prices, and made it a full time job. When he went to a show he swept the board. His rather old fashioned remedy for poultry feeding was to miss one day per week with the food and this gave the birds a sharper appetite; the birds were always fit and strong so they did not suffer. However, he was an expert and knew what he was doing and it is not a recommended practice for amateurs who are not able to assess whether the stock is being fed too much. The size of the crop is a guide, but is not infallible, because a too bulging crop can be an indication that the bird is eating too much and not producing eggs.

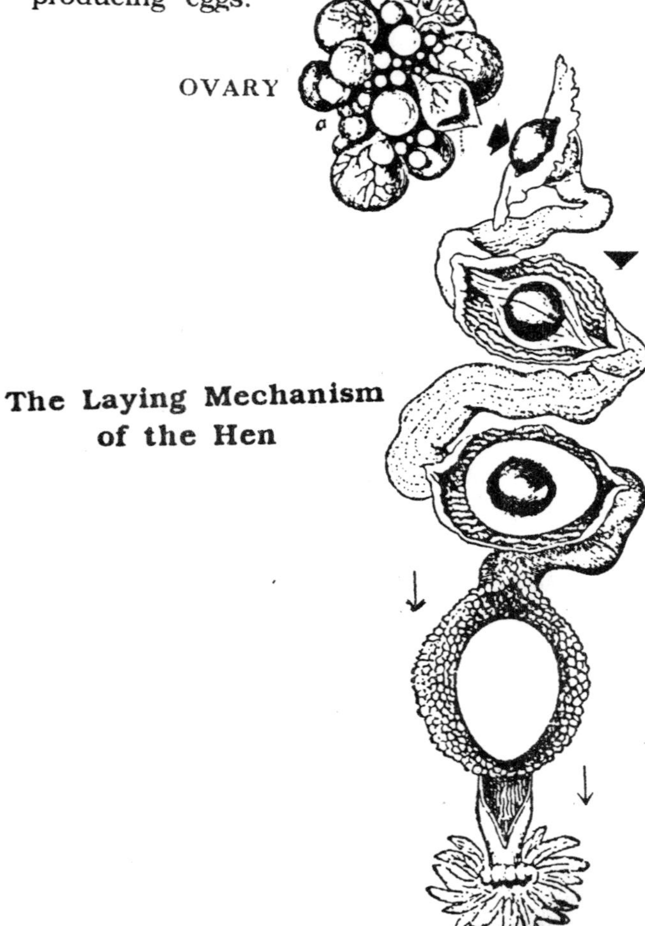

The Laying Mechanism of the Hen

UNDERSTANDING THE HEN

The complex structure of the hen, including the production of eggs, requires special attention to the needs of the birds, thus maximizing production and minimizing ailments. Eggs are developed in the ovary in the form of yolks, rather like a bunch of grapes of varying sizes, the largest being the next to be laid. The final production of each egg depends on the treatment received by the birds.

Once the yolk has reached maturity it detaches itself from the **ovary** and falls into the **oviduct (a long tube)**, and begins its long passage down the tube being given the white of the egg, then the enveloping skins and the shell, including its colouring. All this takes 18 hours or so, and if a hen lays every day it is a continuous process. The egg is laid small end first and is "soft" shelled when it emerges, but hardens very quickly with exposure to the air. If an egg is really soft shelled, ie, no shell at all, then there is a problem, possibly because the hen is too fat and this presses on the lime-secreting glands inside the oviduct and causes them to malfunction.

Matters Which Require Attention for A Contented Hen

The contented hen lays eggs; the troubled hen does not and therefore it is necessary to understand the essential requirements for her welfare.

1. **Airy House (sound ventilation)***
2. **Baths for dusting/cleaning themselves.**
3. **Clean, fresh Water.**
4. **Dry Runs & Litter**
5. **Exact Meal Times.**
6. **Food Made Crumbly if Mash is Fed.**
7. **Greens & Grit regularly.**
8. **Hygiene of a High Standard.**
9. **Indoor Protection in Inclement Weather.**
10. **Jamming (overcrowding) to be avoided.**

Things to Avoid (Trouble Shooting)

These are matters which could affect egg production and therefore must be avoided at all costs. If hens suddenly stop laying there must be a reason. Possibilities are:

1. **Noises - especially sudden and loud.**
2. **Handled roughly and carried by the feet.**

If essential, the correct way is to catch a hen with a large net (usually a fishing net will do) and then tuck her under one arm,

---

*** Not having sweat glands the hens have to dispose of unwanted matter through the digestive system or the lungs, hence the need for adequate fresh air.**

holding her firmly and gently with the arm and steadying with the other hand. Never chase her round and round to catch her. If hens are to be **checked** for health reasons or whether laying (pelvic bone test), this is best done in the evening when they have gone to perch.

3. Draughts and dampness on the floor of the house.

Both these are disliked by birds and will be unproductive.

4. Empty Water Fountains

If hens are left without water they cannot lay or function properly.

5. Vermin

Rats, mice, foxes, cats and dogs are all predators and must be watched.

6. Stale Ground.

7. Lack of Greens.

8. Exposed to Infection.

At the first sign of health problems try to spot the cause and isolate any infected birds. Do not worry about the moult (annual loss of feathers), but check for mite if feathers show bare patches at other times.

9. Dirty Houses & Litter.

Above all, keep the hens interested and active, giving titbits and attention. If treated with affection they will respond; a happy hen is usually a productive hen!

INDEX